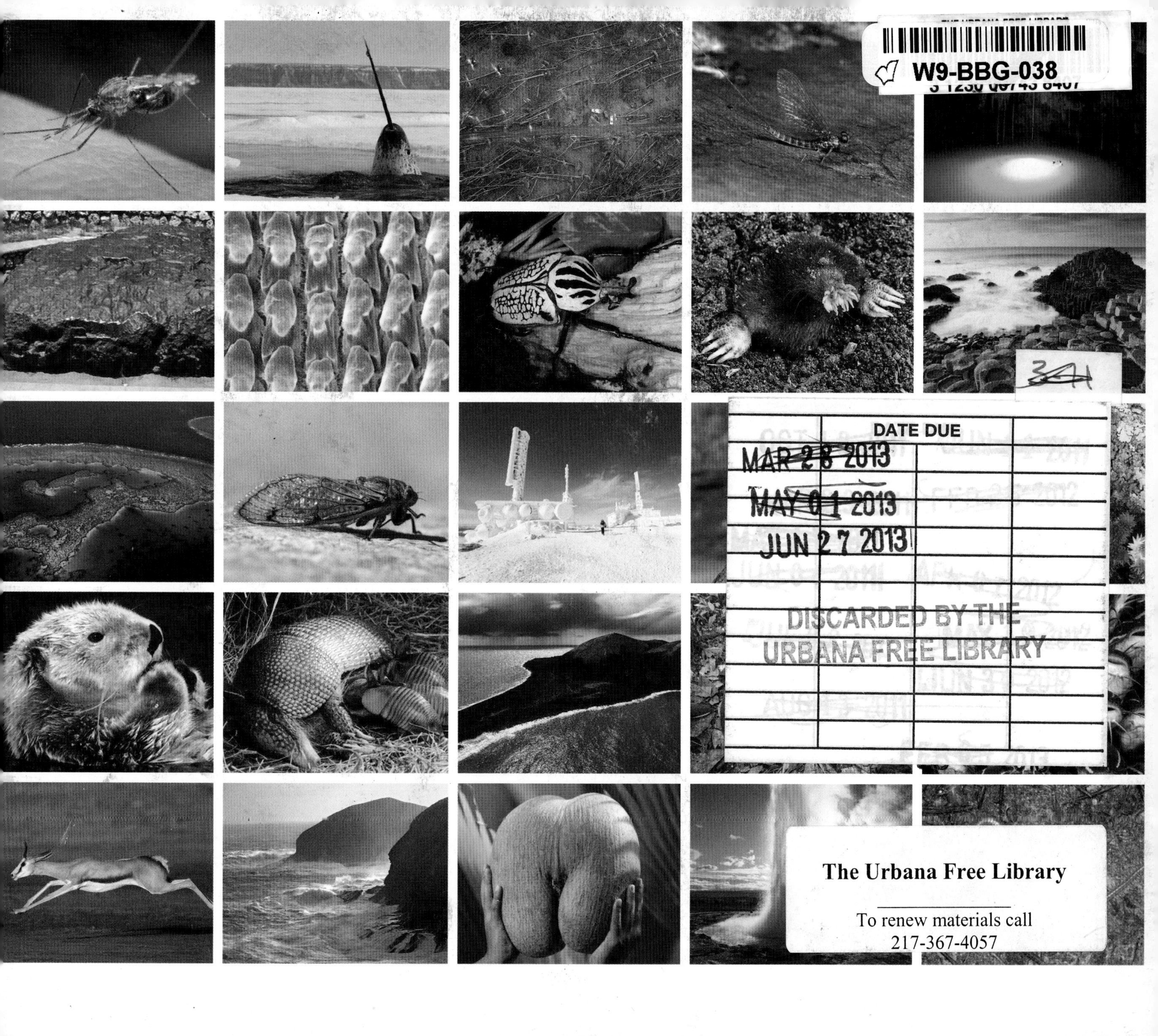

# WORLD OF WONDERS

## The Most Mesmerizing Natural Phenomena on Earth

Library of Congress Cataloging-in-Publication Data

Roman, Elisabeth, 1967–
[Ma terre de tous les records. English]
World of wonders: the most mesmerizing natural phenomena
on Earth / Elisabeth Roman.
p. cm.
ISBN 978-0-8109-8963-4
1. Earth sciences—Miscellanea. 2. Zoology—Miscellanea. 3. Earth sciences—
Miscellanea—Pictorial works. 4. Zoology—
Miscellanea—Pictorial works. I. Title.
QE53.R6613 2010
508—dc22
2009052203

Originally published in French in 2009 by Éditions de la Martinière

English translation by Amanda Katz
Book design by Elisabeth Ferté

Published in 2010 by Abrams Books for Young Readers, an imprint of ABRAMS.


Printed and bound in Singapore
10 9 8 7 6 5 4 3 2 1

115 West 18th Street
New York, NY 10011
www.abramsbooks.com

# WORLD OF WONDERS

## The Most Mesmerizing Natural Phenomena on Earth

ELISABETH ROMAN

Calling all world champions! Not all world records are forged by athletes in stadiums. The earth and its creatures are just as capable of extraordinary feats. What mammal lives at the highest altitude? Where can you breathe the purest air? What tree is the most gigantic? What animal is as noisy as a jackhammer?

In this book, you'll discover some of nature's greatest achievements, along with eye-popping photos. You'll encounter such astonishing sights as a mushroom as tall as a building, lava that flows at sixty miles per hour, and an insect that glows as brightly as a flashlight.

There is truly no end to the earth's incredible accomplishments. No planet could be worthier of its medals.

**Elisabeth Roman**

# Who can see in 3-D?

> The mantis shrimp.

What incredible eyes! Each one grows at the end of a stalk that can pivot independently. This allows the mantis shrimp to see to its left and its right at the same time. It can also see in three dimensions, giving it the ability to spot transparent animals on the seafloor—which it quickly scoops up for lunch.

# What are those green lights in the sky?

> The aurora borealis.

On some nights near the North Pole, particles thrown off by the sun make the sky glow with green light. This spectacular sight is called the aurora borealis, or the northern lights. The most brilliant aurora borealis took place on September 2, 1859: It was visible all the way down to the equator and even farther south—in Australia.

# What grows eight to ten millimeters every year?

> The mountain Nanga Parbat.

This may seem like a very small number, but Nanga Parbat, in the Himalayas, sets the record for the mountain that grows the most each year. Why is it getting taller? In this part of Asia, the surface of the earth is composed of two gigantic tectonic plates, one of which is sliding under the other. Each time the plates move, Nanga Parbat is pushed upward. The Alps are also getting taller—but only by one or two millimeters a year.

# What can fly for the longest time?

> The black swift.

These birds can fly for up to two years without landing. They do everything while in the air: eat, mate, and even sleep. They land very rarely—only to make their nests and sit on their eggs. In fact, their claws are too small for them to walk easily. No wonder they would rather fly!

# How deep is this cave?

> 7,185 feet below ground.

It took centuries of wear by water to dig a cave this deep. The Krubera cave, in the country of Georgia, is the deepest in the world, comparable to an underground parking garage 730 stories deep. People have known about the cave for decades, but in 2001, speleologists, or cave explorers, reached a depth of 5,600 feet, which made it the deepest known cave. They have since gone even deeper, to 7,185 feet. And who knows? There could be other, even deeper caves that we have yet to discover.

# What is the heaviest flower?

> The rafflesia.

The rafflesia flower, which grows in the humid forests of Southeast Asia, measures more than three feet across and weighs up to twenty-two pounds. It's so heavy that it grows right on the ground—no stem could support its weight. Another distinctive trait: It smells vile. Its horrible stench attracts flies, which gather its pollen and carry it to another rafflesia, allowing the two flowers to cross-pollinate and new ones to grow.

# The longest eclipse?

> Seven minutes and twenty-nine seconds.

A long time from now, on July 16, 2186, the inhabitants of our planet will witness a solar eclipse: The moon will pass between the sun and the earth, hiding the sun completely. When this happens, it will get as dark as night in the middle of the day. The eclipse of 2186 will be the longest one in history.

# How many wooden matches could you make with this tree?

> Five billion.

This giant sequoia, called the General Sherman, grows in California and holds the record for the highest-volume tree in the world. It's 275 feet tall (more than twenty-eight stories) and more than thirty-six feet across (as wide as three highway lanes). With its wood, you could build forty houses with five rooms each. But not so fast: This tree is protected, and it's against the law to cut it down.

# Which sea is the warmest?

> The Red Sea.

This sea is as warm as a swimming pool. The Red Sea, in the Middle East, is the warmest in the world: Its surface temperature stays between 70 and 77 degrees Fahrenheit all year round, even in winter. Temperatures have even been recorded as high as 89 degrees Fahrenheit. This heat, produced by underwater volcanic activity and the deserts that surround the sea, attracts more than a thousand species of fish of every variety and color.

# What is the coldest recorded temperature?

> −128.6 degrees Fahrenheit.

On July 21, 1983, while some people went to the beach, others turned up the heat, including the scientists at the Russian base of Vostok, at the heart of the South Pole. That day, the thermometer there registered a temperature of −128.6 degrees Fahrenheit. But at Vostok, everyone wears a parka year-round anyway: The temperature never rises above 14 degrees Fahrenheit.

# Where is it night almost all the time?

> In the South Orkney Islands.

Welcome to the South Orkney Islands, near the South Pole. Conditions are so harsh that only around twenty scientists live there. The worst part—besides the cold and the violent winds—is the darkness. The clouds are so heavy that the night lasts on average around twenty-three hours a day.

# What built these ancient totem poles?

> Water, wind, and ice.

These elements worked together for forty million years to create this magical landscape. The Bryce Canyon area, in Utah, is home to the largest number of these astonishing natural pillars—thousands of them. They have a funny name: hoodoos, or fairy chimneys. But you are unlikely to see any sorcery there, just squirrels darting up and down the rocks.

# Who's making that bright shining light?

> The fire beetle.

With two green lights on its thorax and an orange one on its stomach, the fire beetle, or cucujo (pronounced *koo-KOO-ho*), is the world's most luminous insect. Forty cucujos create as much light as a candle. They are native to Mexico and the West Indies, where at one time people used them for light at night and even made them into jewelry. Women used to slip them into bags of transparent cloth and make fiery cucujo necklaces and brooches.

# What animal is slower than a snail?

> The sloth, which takes one minute to go ten feet.

Here is an animal that truly deserves its name. The sloth, which lives in Central America, is the slowest of the mammals. Its slowness makes it vulnerable to its predators: the jaguar, the ocelot, and the eagle. So it prefers to live suspended from the branches of trees, where it finds its food. It only comes down to the ground to leave its droppings three times a month.

# How fast is this lava flowing?

> At sixty miles an hour.

What amazing speed! No one has ever seen lava rush out of a volcano so fast. Usually, a lava stream flows around twelve and a half miles an hour. This scene played out on January 10, 1977, in Congo, Africa. On that day, a flank of the volcano Nyiragongo fissured, and the whole lava lake drained out of the crater in less than an hour.

# How tall is this enormous wave?

> As much as twenty-nine feet tall.

Watch out—this wave is nicknamed the Silver Dragon. It's called *silver* because of its gray-blue water and *dragon* because of its enormous size. This wave can be up to three stories tall, or nearly thirty feet in height. Each fall, at the moment of the highest tide, the gigantic Silver Dragon sweeps back up the Quiantang River in China from the ocean. It makes so much noise that it can be heard an hour before its arrival. This strange phenomenon is known as a tidal bore. Other rivers in the world, like the Gironde in France, also experience tidal bores, but far less impressive ones.

# What animal wins the medal for the long jump?

> The springbok, an African antelope.

The springbok's leaps can be as long as fifty feet. But this graceful mammal has another athletic talent: It can jump vertically as well, up to ten feet in the air—like a spring. What's the point of this skill? No one knows exactly, but the springbok may be sending a message to lions and leopards, its predators: "Don't bother chasing me—I'm too fast and agile to catch!"

# Which was the most destructive hurricane?

> Hurricane Katrina.

Katrina, which struck New Orleans, Louisiana, on August 29, 2005, is the most catastrophic hurricane ever recorded, with an eye (a center) almost twenty-five miles in diameter and winds of up to 174 miles per hour. The cost of the damage approached $110 billion.

# What animal's heart beats the fastest?

> The hummingbird's, at 1,260 beats a minute.

The blue-throated hummingbird's heart beats faster than any other animal's. By comparison, the human heart beats only seventy times a minute. Why this great velocity? The hummingbird must flap its wings rapidly in order to hover and sip nectar from flowers. And the only way to do this is for its heart to beat quickly, too.

# What lays the most eggs?

> The ocean sunfish.

The bigger the fish, the more eggs they produce: This is the distinctive quality of sunfish, which are found in all warm and temperate seas. So a sunfish five feet long produces around three hundred thousand eggs at a time, and a sunfish ten feet long can produce one billion eggs.

# How wide is the biggest leaf?

> 9.8 feet across.

The *Victoria amazonica*, a South American plant, makes a striking impression with its huge floating leaves—strong enough to support the weight of a forty-four-pound child, like a natural boat. In the summer, the plant offers another beautiful sight, although all too briefly: Its big white flowers last only two days before fading.

# The most piercing eyesight?

> The eagle, which can see nearly a mile away.

While it glides at high altitude, the eagle scans the ground looking for food. It has the ability to spot its prey from far above, even an animal as small as a mouse. The eagle owes its exceptional eyesight—2.5 times better than that of humans—to the size, shape, and position of its eyes. In particular, the eagle's retinas have one million light-sensitive cells per square millimeter (versus two hundred thousand for humans), which makes its vision much sharper.

# What animal is the smelliest?

> The skunk: You can smell it up to 1.9 miles away.

They may look adorable, but these two striped skunks are hiding a secret. If they sense danger, they spray their enemy with a disgusting oily yellow liquid that comes from glands just under their tails. Skunks can spray up to twenty feet away. The awful stink they create can be detected for up to 1.9 miles, the record for the smell that reaches the farthest distance. It is also very difficult to get rid of.

# What is the most active volcano?

**> Kilauea: Its eruptions can last for decades.**

The volcano Kilauea, on the island of Hawaii, spews lava all the time. It started on January 3, 1983, and hasn't stopped since. The lava flows have destroyed several villages, hundreds of buildings, and miles of road, and it is therefore considered the most active volcano in the world. But there are others that give it some competition, such as Piton de la Fournaise, on the island of Réunion. That volcano has erupted every year for the last century.

# What weaves a giant web?

> The tropical spider *Nephila*.

The *Nephila* does things on a big scale. This giant spider (up to four inches across) weaves webs in just a few hours that can be twenty feet tall by six and a half feet wide. This size is useful for trapping a variety of big insects. In fact, the silk threads are so strong, even small birds get their feet caught in them. But the *Nephila* will not eat those!

# Why should you avoid this flower ?

> Because it smells awful!

Not all flowers smell good. The arum titan, which grows in the wild only on the island of Sumatra, holds two records: the tallest flower and the one that stinks the most. It can reach thirteen feet in height, more than one story tall. And its horrendous scent can be detected up to six-tenths of a mile away.

# What is the most dangerous animal?

> The mosquito.

Every year, the mosquito (a different kind from the one that bites us in the summer) kills between one and two million people in Asia and Africa. There are 3,500 species of mosquitoes, and 150 of them often carry fatal diseases, such as malaria and yellow fever, that can be transmitted to humans through one little bite.

# What is this horn for?

> It's not a horn; it's a ten-foot-long tooth.

The narwhal's horn is not a true horn, but a tooth, which can be up to ten feet long. It is commonly thought that the narwhal uses it for hunting. In fact, another explanation has recently emerged. Through tiny receptors, it provides the narwhal with a wealth of information about the water surrounding it, such as temperature and saltiness.

# The storm of the century?

## > It occurred in Europe in 1999.

No storm in the world has uprooted so many trees. On December 26 and 27, 1999, Europe (and France in particular) was swept from west to east by two terrible hurricanes named Lothar and Martin. The winds blew up to 162 miles an hour and brought down 350 million trees. It took years to remove all the uprooted trees and clear up the forests.

# The shortest life span?

## > Just a few hours long.

The mayfly belongs to an order of insects called Ephemeroptera, which is a good name for them. *Ephemeral* means "lasting a very short time," and the life of the mayfly is short indeed. It spends a year as a larva and then takes its adult form—but only for a few hours, just long enough to find a mate and reproduce. Nature has not even given it a mouth: It dies so quickly that it doesn't get a chance to eat.

# What animal swims the fastest?

> The Atlantic sailfish.

Folded on the sailfish's back is a fin in the form of a sail (hence its name) that allows it to swim faster. When it wants to zoom ahead, it unfurls its sail and swims at top speed, up to seventy miles an hour. This fish, which lives in the Atlantic Ocean, is the fastest in the world. It's also a bruiser: It can measure up to ten feet long and weigh up to eighty-eight pounds.

# How much does the biggest meteorite weigh?

> More than sixty tons.

This giant rock from space landed in the African country of Namibia. It weighs sixty tons, equal to the weight of ten elephants. This meteorite landed eighty thousand years ago but was not discovered until 1920, by a farmer working in his field.

# What has the biggest seed?

The sea coconut. It weighs forty-four pounds

The coco de mer, or sea coconut, which is in the palm tree family, grows on only two islands in the world: Praslin and Curieuse, in the Indian Ocean. Its giant double seeds are so heavy that they can't be transported by either wind or animals. For this reason, the trees always grow in the same place.

# What is the heavyweight of the insect kingdom?

> The Goliath beetle, found in the African forest.

The Goliath can measure up to 4.3 inches long and weigh up to 3.5 ounces, as much as a mouse. Its wings beat so quickly when it flies that it makes a noise like a helicopter. Despite what its huge size might lead you to believe, the Goliath is not at all dangerous to humans.

# How high can lightning reach in the sky?

> Around sixty miles up.

In the 1990s, it was discovered that during certain storms, strange flashes of light appear underneath high-altitude clouds, around sixty miles up. Scientists gave them names: The red ones are called sprites, the blue cones are called jets, and the ones shaped like a disk are called elves. But no one knows what causes these high-altitude flashes of light, the highest lightning that exists.

# How far does this desert extend?

> More than 250,000 square miles.

The Sahara is the biggest desert in the world, but it is only partly composed of sand. Holding the title for the largest sand desert, at more than 250,000 square miles, is the desert of Rub'al-Khali, which is located partly in Saudi Arabia. In certain places, the dunes reach higher than 650 feet, as much as a seventy-story building. It rains so little that only around twenty species of plants and a few spiders and rodents are able to live there.

# What is the most prickly animal?

> The porcupine.

Be careful not to approach the cape porcupine, an African species, too closely. Its thirty thousand spines can be up to twenty inches long. And the porcupine uses them as a weapon. When it is attacked by a bear, a fox, or an eagle, it unsheathes its many swords—and often succeeds in killing its predators.

# How often do earthquakes happen in Japan?

> Every twelve days.

In some regions, earthquakes are extremely common. This is the case in Japan, which has been shaken by one thousand violent quakes since 1973. On average, that's one every twelve days. Fortunately, the Japanese are well prepared for these catastrophic events. Every year on September 1, the whole country holds a giant safety drill to remind everyone what to do in case of an earthquake.

坂田鍼灸

# How fast can this slithery snake go?

> More than eighteen miles an hour.

This snake, the black mamba, is quite the champion. For starters, it holds the title of the longest snake in Africa: It can grow to more than eight feet long. But it takes another trophy for being the fastest snake in the world. At twelve miles an hour, it slithers as fast as a man running. When it goes into a sprint, it can go even faster: more than eighteen miles an hour.

# What is hiding inside this female sea worm?

> The male sea worm.

She's three feet long; he's only one to three millimeters! The green spoon worm, a kind of sea worm, wins the award for the couple with the greatest difference in size. The male green spoon worm lives its whole life inside or clinging to the female's body in order for her to reproduce. And there may be as many as eighty-five males inside the female.

# What's the oldest kind of fish?

> The coelacanth: it lived alongside the dinosaurs.

It was believed the coelacanth appeared on earth 390 million years ago and disappeared at the same time as the dinosaurs, 65 million years ago—but no. In 1938, a fishing boat caught one in the Indian Ocean, near South Africa. About two hundred more were captured after that. But it is now forbidden to catch this prehistoric fish, of which so few remain.

# Which animal is big enough to crush the scale?

> On land, the African elephant.

The African elephant is a proud member of the heavyweight team. At seven and a half tons and thirteen feet tall, it is the biggest living land animal. As a result of its huge size, it must eat gigantic meals: Each day it swallows between 200 and 650 pounds of food and drinks fifty gallons of water, enough to fill a bathtub.

# What grows the fastest?

> The giant bamboo plant.

The giant bamboo plant truly deserves its name. This bamboo variety grows faster than any other plant. It shoots up by more than three feet a day and takes only a few weeks to reach its full size: more than sixty-five feet tall and twenty-three inches in circumference. It's a gigantic size for a blade of grass. (Yes, bamboo is not actually a tree but a kind of grass.)

# What animal can double its weight in four days?

> The hooded seal.

What an adorable baby! At birth, the hooded seal, which lives in the ocean near the North Pole, already weighs 55 pounds. Four days later, the seal calf's weight has doubled to 110 pounds. Certainly its mother's milk, of which it drinks two and a half gallons every day, is very nourishing: It contains fifteen times more fat than cow's milk.

# What is the tallest tree in the world?

> This sequoia, which is 379.1 feet tall.

This champion was discovered in 2006 in a national park in northern California. But the exact location of the seven-hundred-year-old sequoia has not been revealed, for fear that thousands of tourists would come to see it and harm its environment. Named Hyperion, after a Greek god, this tree might be even a little taller if it wasn't for damage caused by woodpeckers at the top.

# The highest cliff?

> 4,100 feet tall.

As tall as four Eiffel Towers stacked on top of one another, this is the tallest purely vertical cliff face in the world, on Mount Thor on Baffin Island in Canada. Daredevils come from around the world to experience its thrills. They leap from the top and spend thirty seconds in free fall before opening their parachutes.

# What is nicknamed "the roof of the world"?

> Mount Everest, which is 29,035 feet tall.

The Tibetans call it Chomolungma ("Goddess Mother of the World"), and the Nepalese call it Sagarmatha ("Forehead of the Sky"). For everyone, it is the summit of the world: the tallest mountain, the one that mountaineers dream of scaling. The first successful ascent was in May 1953. In the years since, Everest has been climbed nearly four thousand times. The record for speed? Ten hours and fifty-six minutes to the summit.

# How high is this vulture gliding?

> 36,000 feet up, as high as an airplane.

On November 29, 1973, a terrible collision took place. A Rüppell's vulture crashed into an airplane flying 36,000 feet over Africa and got caught in the engine. To reach this altitude, the bird was lifted by an air current. No other bird has ever been seen so high. Another champion of the sky is the goose, which can climb to 29,500 feet.

# What mammal lives at the highest altitude?

> The pika.

This relative of the rabbit is not afraid of cold or heights: It lives at an altitude of twenty thousand feet on Mount Everest, in Asia. It digs burrows into the side of the mountain, in which it can take shelter and store its food. Even tougher and higher up is the jumping spider, which has been found at twenty-two thousand feet. But at this height, there are no plants or other insects, so it has to come downhill to find its dinner.

# The deepest place on earth?

> An underwater pit 35,797 feet below sea level.

This spot is known as the Challenger Deep, in the Pacific Ocean, and it is the deepest point on the planet. We owe this precise measurement to a small Japanese robot that descended to the bottom in 1995. To the scientists' surprise, the robot even brought back microscopic organisms from its voyage. It had been thought the pressure was so great at such depths that no life was possible.

# Is this just a simple lake?

**> No, it's the earth's deepest natural well.**

In this region of central Mexico, there are many natural wells, called *cenotes* (pronounced *see-NOH-tays*). Cenotes are sinkholes in the earth, not wide but extremely deep and full of water. The deepest, the Zacatón, is 1,100 feet deep. The water is fresh and warm—86 degrees Fahrenheit, the perfect temperature for a bath. The reason for this heat: subterranean volcanic activity.

# Is this just a maze of trees?

## > No, it's the biggest forest in the world.

Pines, firs, and spruce trees as far as the eye can see . . . with its 6.5 million square miles, the taiga is the world's largest forest. Better not get lost! The taiga, which forms a ring around the North Pole, represents a third of the forest on earth. About half of the taiga is located in northern Russia; the other half is in Alaska, Canada, and Scandinavia.

# What is this strange white expanse?

> A salt desert, in Bolivia.

In the salt flats of Uyuni, salt stretches as far as you can see. This is the largest salt desert in the world, one hundred times the size of Paris. Forty thousand years ago, this spot held a vast and shallow salt lake, Lake Minchin, which gradually dried up. Only the salt remains.

# What on earth is this?

> Lakes and islands, as seen from the sky.

Look closely: In the center of the picture, the tiny circle of water is the world's largest lake on an island that is itself on a lake in the middle of an island. Did you follow that? Trace the photo with your finger, and we'll start again from the outside. In the center of the big island Luzon, in the Philippines, is Taal Lake, in the middle of which lies Volcano Island, still an active volcano. The crater of the volcano holds another lake, called Crater Lake. It is the most famous lake on an island on a lake on an island in the world.

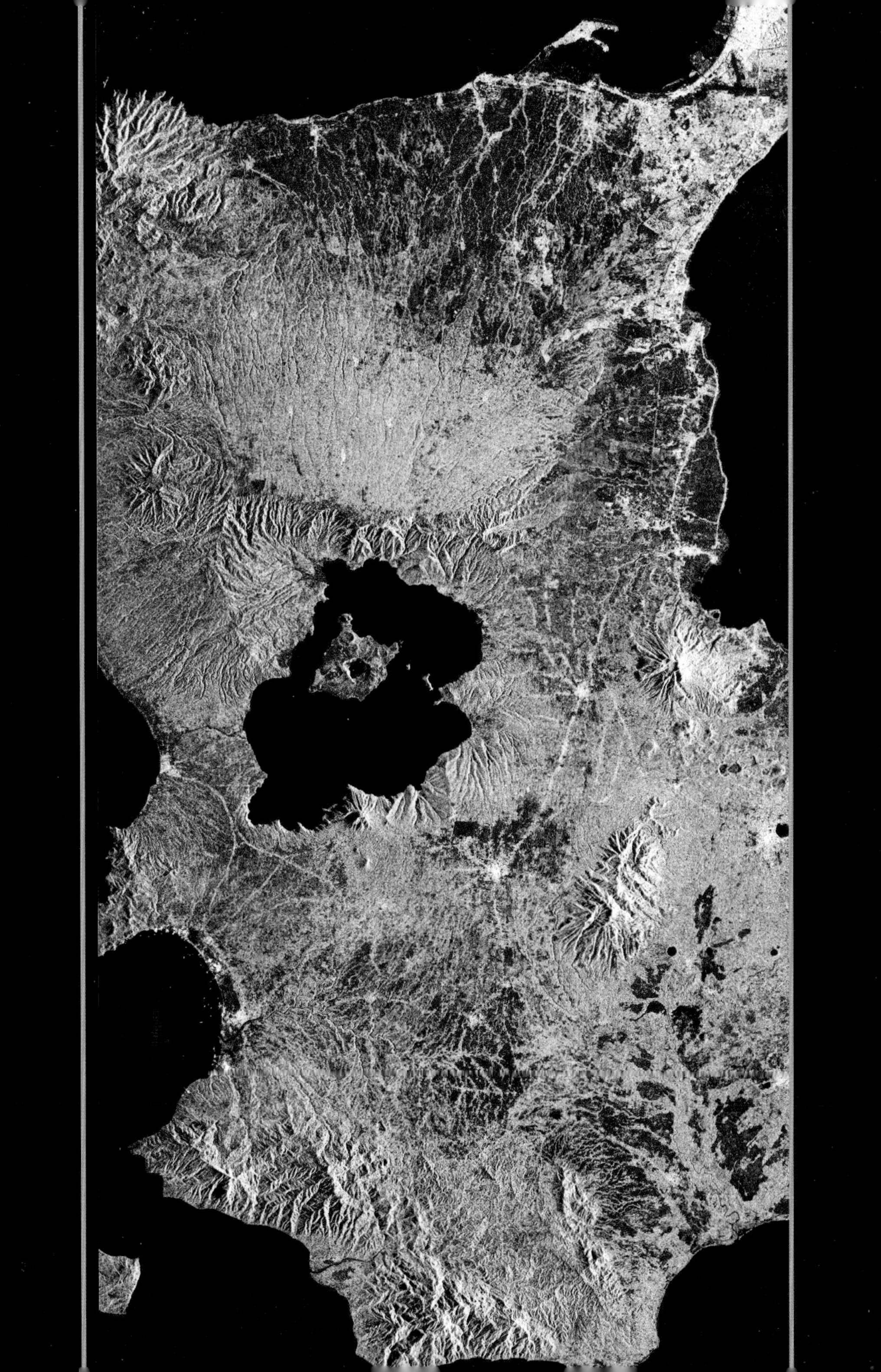

# Is there such a thing as quicksand that moves?

> Yes, in England.

Watch where you put your feet! When the Irish Sea pulls back, at low tide, the vastest expanse of quicksand in the world (120 square miles) appears in Morecambe Bay, in England. The sands can swallow those who walk across them. But this expanse is not always situated in the same place. There soon will be a satellite that observes the area from space and provides live information about the whereabouts of these dangerous sands.

# What plant has the longest hooks?

> The devil's claw.

Certain plants come equipped with hooks, and the devil's claw has the longest: up to 13.8 inches long. But what for? To catch at the feet of animals as they go by. This way, the devil's claw is able to travel and to disperse its seeds over long distances.

# What is this squirrel doing?

> It's gliding 1,476 feet.

The giant flying squirrel of Asia, related to the common American squirrel, unfurls the membranes connecting its front and back legs, launches itself from a branch—and glides! It can even steer, thanks to its tail, which serves as a rudder. A tilt to the left, a tilt to the right, and it can soar right through the branches.

# Is this a group of islands?

> No, it's the world's largest animal.

We are looking at the Great Barrier Reef, on the east coast of Australia. The coral is a minuscule ocean animal that by itself is invisible. But it lives with billions of its fellows, and each one constructs a kind of shell that protects and connects them. Over the course of centuries, the shells thicken and become an enormous stony mass. This reef, the largest in the world, is 1,615 miles long, and its surface measures 135,135 square miles, as big as Germany.

# What is the biggest cavern in the world?

> Mammoth Cave, in Kentucky.

If you walked without stopping, night and day, it would take nearly eight days to walk the 367 miles of galleries that make up Mammoth Cave, so called because of its immense size. On the way, despite the darkness, you would pass thousands of species of plants and more than 130 species of animals. The cavern's subterranean lakes, for example, are home to fish without eyes. Because it is night all the time underground, the fish have no need for them.

# How big are the crystals in this cave?

> Up to forty-six feet long.

In this cave, the temperature stays around 122 degrees Fahrenheit. It's impossible to stay there for longer than a few minutes. But this is exactly the temperature and humidity that allow these crystals of the mineral gypsum, in the Naica Mine in Mexico, to grow so big. The longest one measures forty-six feet.

# How long is the longest fjord in the world?

> It extends for 217 miles.

A mountain of ice to the right; a mountain of ice to the left. This is the majestic landscape confronted by those who navigate the salt waters of the valley of ice known as a fjord. Scoresby Sund, to the east of Greenland, holds the record for the longest fjord in the world; it extends for 217 miles. Where did it get this name? In 1822, the whale hunter William Scoresby was the first to explore it.

# Where is the greatest cache of crystals?

> In Almería, Spain.

A geode is a hollow stone whose center is spiky with crystals. Most are small enough to hold in your hand. This geode, in Almería, Spain, is gigantic: twenty-six feet long, six feet wide, and six feet high—big enough to hold ten people. Not so easy to slip it in your pocket!

# What animal can endure the most extreme conditions?

> The water bear.

Minuscule (no larger than 0.06 inches) but incredibly resistant to extremes, the tardigrade, or water bear, can adapt to nearly any condition. It can survive for several minutes in an oven at 300 degrees Fahrenheit, nearly a day in extreme cold (–460 degrees Fahrenheit), and even in outer space! In addition, it is found all over the planet: on top of the Himalayas, in the deepest of oceans, at the poles, and at the equator.

# Just a little whirlpool?

> No, it's the most powerful current in the ocean.

Boats must navigate carefully through this spot in Norway, on the Saltstraumen Sound. Every six hours, as the tides change, millions of gallons of salt water arrive at the speed of twenty-five miles an hour. The current is so powerful and the passage so narrow, it creates a giant whirlpool thirty-three feet across. An amazing spectacle—and one that might make you dizzy.

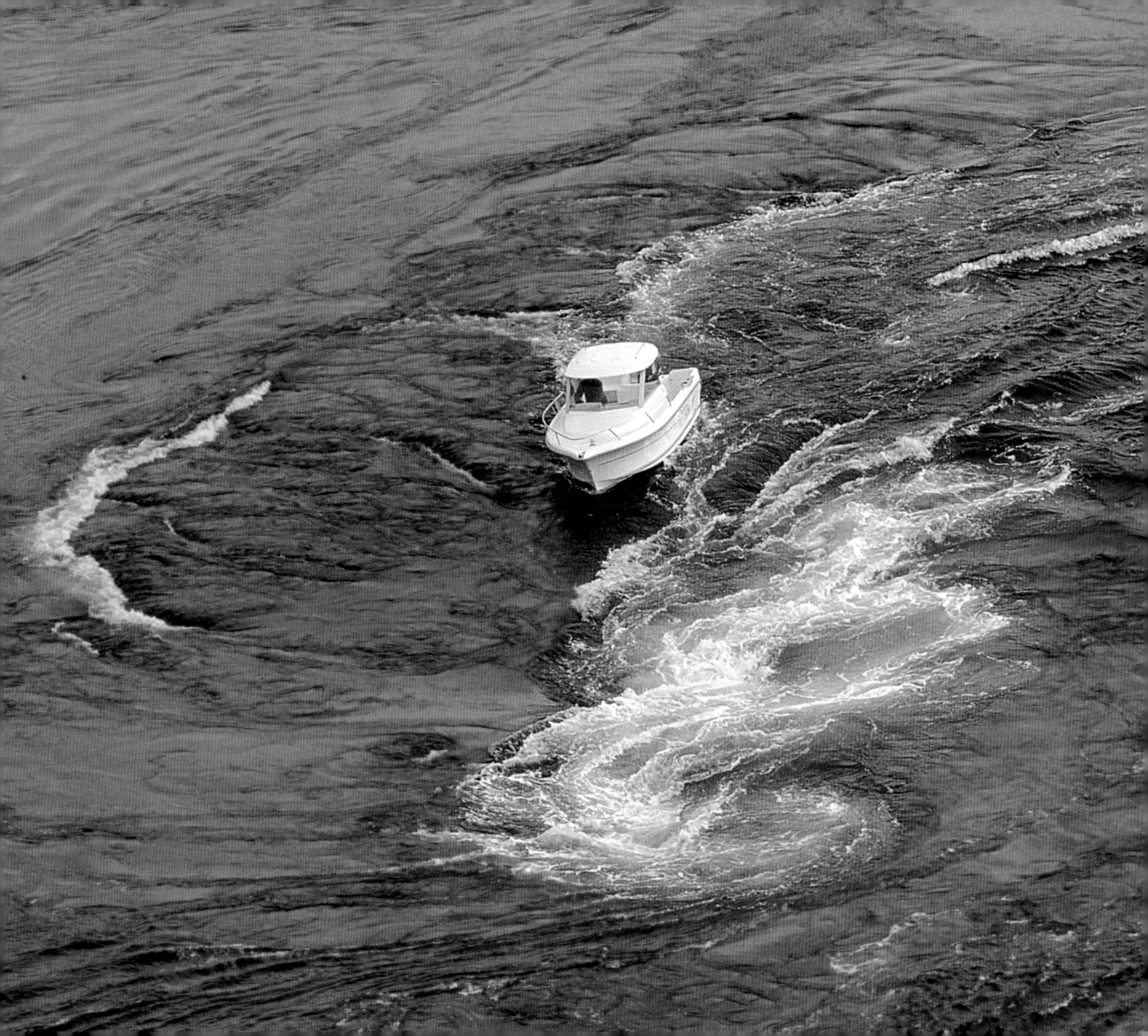

# What wind is the strongest?

> The wind on top of Mount Washington.

On top of Mount Washington, in the White Mountains of New Hampshire, you might as well not wear a hat, because it would blow away immediately. The fiercest winds in the world blow through this spot, as strong as those in a hurricane: seventy-five miles an hour or more. It was on April 12, 1934, that the strongest of them all was recorded: 231 miles an hour! We can imagine that even more powerful winds have occurred there, but no one has been able to prove it because the instruments used to measure wind speed have not been strong enough to withstand them.

# What animal eats the fastest?

> The star-nosed mole eats in less than a second.

Look closely at the nose of this mole: There are twenty-two tentacles at the end. This is why it's called the star-nosed mole. It is blind, but its tentacles allow it to detect, identify, and swallow its prey in record time—less than a quarter of a second. When you think that a train conductor takes longer to react to danger on the tracks and hit the brakes, the mole's speed seems quite impressive.

# I'm always on time. Who am I?

> The geyser Strokkur, in Iceland.

Every five to ten minutes, for a few seconds, this geyser erupts and shoots hot water up to sixty-six feet in the air. It is the most punctual geyser in the world—unlike its neighbor, the Geysir, which often stays inactive for years at a time. What's the cause of this natural fountain? The Strokkur is located in a volcanic region, and volcanic gases deep below the earth push the water toward the surface.

# What has the greatest number of stairs?

> The Giant's Causeway.

The natural wonder known as the Giant's Causeway, in Ireland, comprises forty thousand pillars ranging from a few inches to thirty-nine feet tall. As you climb down toward the ocean, the columns get smaller, like stairs on a badly built staircase. These columns are sixty million years old, dating to the time when volcanoes were active in the region, and they are made of cooled lava.

# What is this odd pattern?

> A snail's tongue.

Seen through a microscope, the tongue of a snail turns out to be covered with minuscule teeth—up to twenty-seven thousand of them—which allow it to grate and grind its food. The snail's tongue works like a moving sidewalk: When the teeth in front wear down, they are replaced by new teeth situated in the back of the tongue.

# How many bats are there here?

> More than twenty million.

What an enormous gathering of bats: twenty million of them! Each year, for the last ten thousand years, in March or April, the same event occurs. The biggest colony of mammals in the world leaves Mexico in search of food and goes to spend the summer in the Bracken Cave, near San Antonio, Texas. As soon as they arrive, the females give birth to their young. And each night, the mother and father bats leave the cave to hunt. There are so many, it takes them three hours to exit the cave.

# Where can you breathe the purest air?

> In northwest Tasmania.

Breathe in, breathe out! Our planet is so polluted that you may well ask if there's any place the air is truly clean. Well, yes. On Cape Grim, in Tasmania (an island to the south of Australia), you can breathe in peace. Why? On this island, there are few polluting factories, and the winds that sweep across it come directly from the South Pole, which is unspoiled and nearly ten thousand miles away.

# What has the most feathers?

> The penguin.

Penguins are the animals with the greatest number of feathers. And they need them to withstand temperatures at the South Pole, which go down to –76 degrees Fahrenheit. Each square inch of their skin grows around seventy feathers, three to four times as many as most birds. This adds up to a layer of feathers more than four-tenths of an inch thick, and between ten thousand and one hundred thousand feathers altogether, depending on the species. All those feathers must keep them warm indeed!

# What animal is the furriest?

## > The sea otter.

The sea otter spends its days splashing in the icy waters in the northern Pacific Ocean. To do this without shivering, it needs a very thick coat. So whereas a dog's coat has around 650 to 3,900 hairs per square inch, the otter's has a record 903,000 to 1,100,000. Like a wet suit, this fur coat prevents the water from touching the sea otter's skin. And to make it even more watertight, the otter spends many hours a day coating its fur with an oil produced by its body.

# What is the best taster?

> The catfish: It has 100,000 taste buds.

Now here's a fish with taste! The catfish has one hundred thousand taste buds, the organs responsible for recognizing flavors: in its mouth, around its mouth, but also on its barbels (the flesh around its mouth that looks like whiskers). By comparison, we humans have only ten thousand taste buds on our tongues. What are all these taste buds for? The catfish often hunts at night, and because it can't see anything in the dark, the taste buds allow it to consider its prey and decide whether to eat it.

# What is the most dangerous tree?

> The manchineel: It can kill you if you eat it.

People often put a sign on the manchineel tree warning passersby not to come any closer. This tree, which grows in the Arabian Peninsula, near the equator in the Americas, and in the Antilles in the Caribbean, is toxic from top to bottom. If you eat its pear-shaped fruit, you will need to be rushed straight to the hospital. Touching its trunk will give you a serious burn. And when it rains, the water that runs off its leaves is as scorching as acid.

# How many babies does this armadillo have at once?

> Four at a time—and all identical.

The nine-banded armadillo is an unusual creature indeed. To begin with, it is covered with an armor that includes nine articulated parts—hence its name. But what makes this animal unique is the number of babies that the female produces: always four at a time and totally identical. In this way, it is unlike any other mammal.

# What is the youngest landmass?

> The island Surtsey, at only forty-six years old.

On November 10, 1963, in the Atlantic Ocean near the coast of Iceland, a volcanic eruption started 427 feet below the water's surface. Tons of lava surged out of the earth. It built up and hardened under the water, and within a few days it broke the surface and formed a small island. The eruption continued until June 1967. Ever since then, the island has been getting smaller again. Only scientists are allowed to go there to study it.

# What living thing can sleep the longest?

> The rose of Jericho—for fifty years.

Like Sleeping Beauty, this plant, which is native to the deserts of Mexico, can lie dormant for up to fifty years. It may look dead, but give it a little water and a miracle takes place. The dry, shriveled leaves suddenly open up and turn green again. This ability allows the rose of Jericho to survive even the longest droughts.

"Sleeping" rose of Jericho with its leaves shriveled and dry.

"Awake" rose of Jericho with its leaves open.

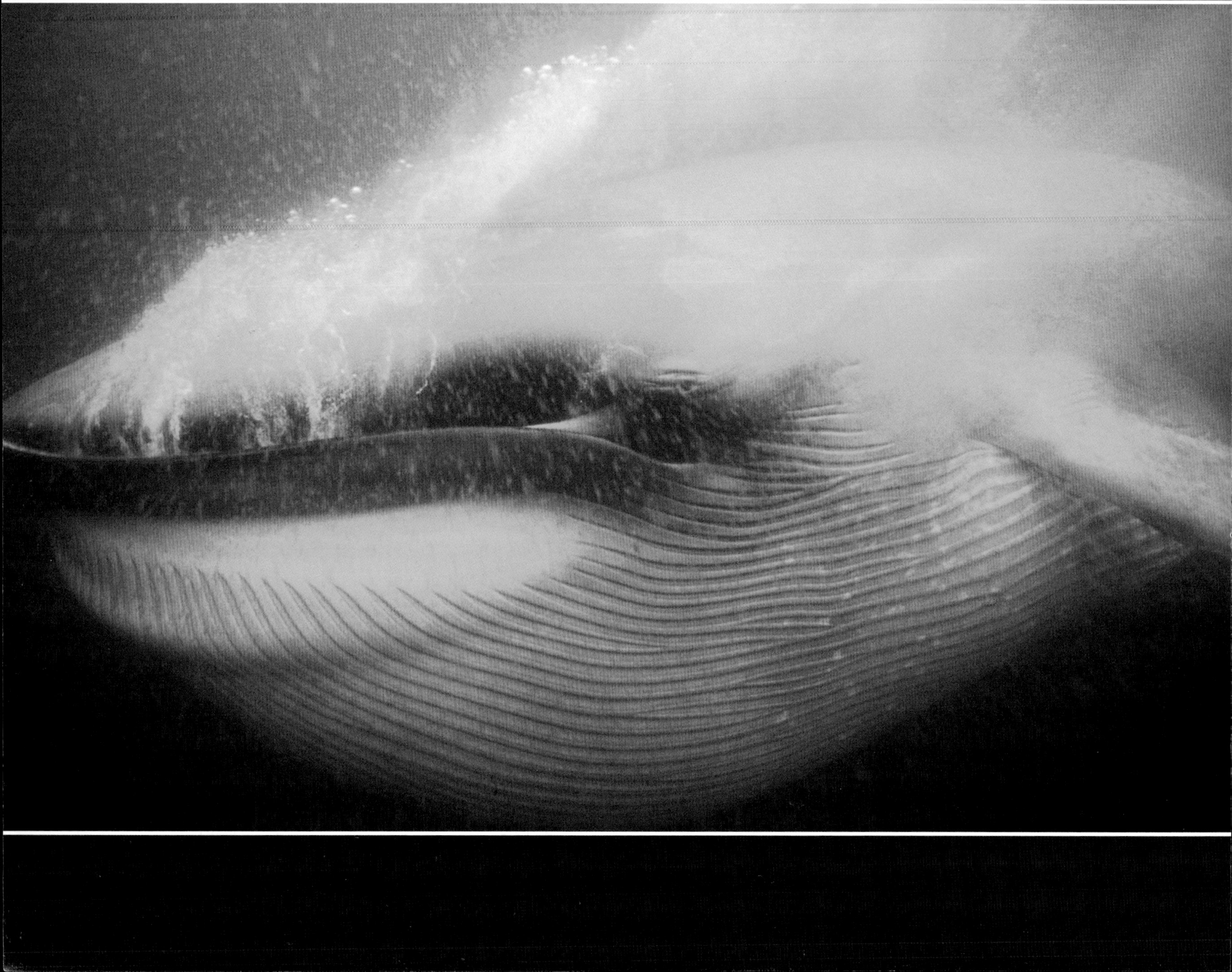

# What is the biggest animal?

> The blue whale.

As long as three buses (ninety-nine feet), as heavy as twenty-six elephants (180 tons), and with a tongue the weight of four cars (four tons), the blue whale is an impressive beast indeed. No land animal can compete. Only the fact that it lives in water allows the whale to support such a heavy body. On land, its bones wouldn't be strong enough and would break!

# How long can this animal go without eating?

> Up to twenty years.

This tiny creature, the tick, eats only three meals during its life. It must wait until a small mammal passes within reach, and then it quickly leaps aboard and sucks the animal's blood. With each meal, the tick gets bigger. At the beginning of its life, when it is still a larva, its first feast allows it to grow into a nymph. With its second, it becomes an adult. It eats its third meal just before reproducing. And because its prey may not pass by every day, the tick must wait for a victim to come along before it can gorge itself on blood. This can take five, ten, or even twenty years. By the time the tick reaches adulthood, it has become one tough bug.

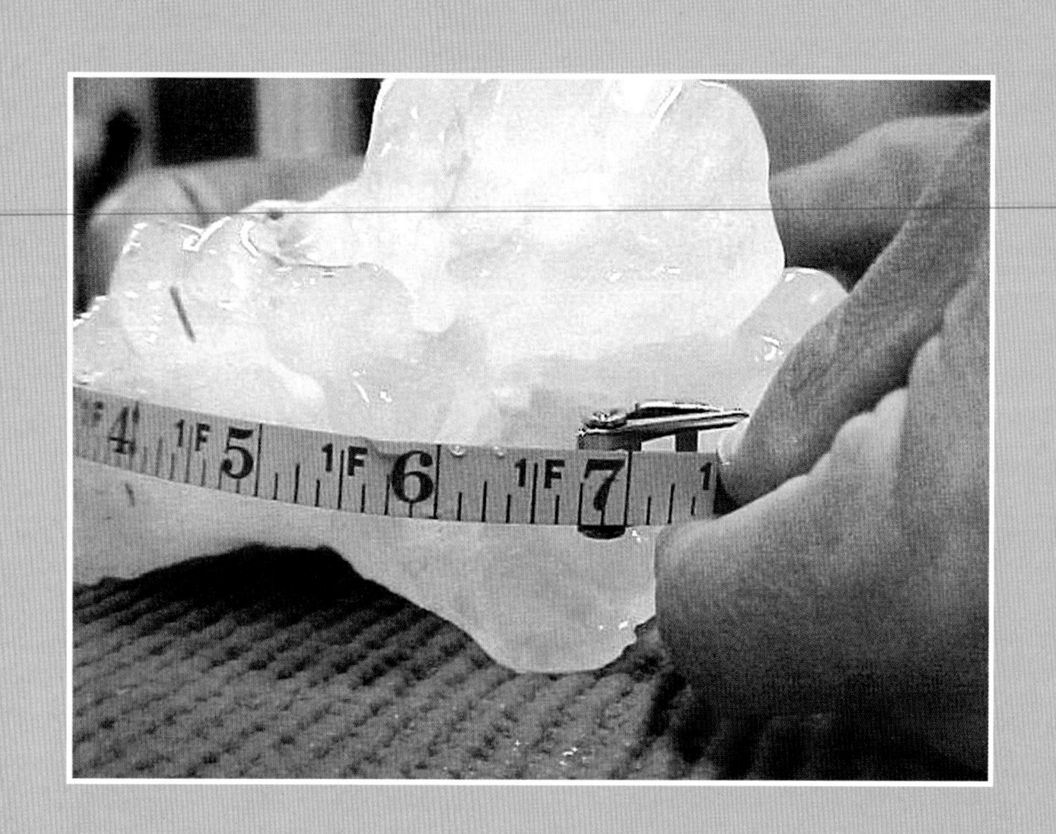
4 1F 5 1F 6 1F 7

# What are as big as melons?

> The largest hailstones.

This enormous 1.3-pound block of ice fell from the sky in Nebraska in 2003. Fortunately, no one was hurt. Unfortunately, we can't say the same for the heaviest hailstones ever recorded. These hailstones, which weighed around 2.2 pounds each and fell in Bangladesh in 1986, killed ninety-two people.

# What has the biggest heart?

> The giraffe.

The heart of the giraffe weighs twenty-four pounds, as much as a small dog. The giraffe is the land mammal with the heaviest heart in relation to its body. Why is it so big? Because the giraffe is huge. The giraffe's blood needs to flow all the way up to its brain, in a head perched at the end of a very long neck. There's only one way to get it up there: an extremely powerful heart.

# What is the hungriest animal?

> The Eurasian pygmy shrew.

Can you believe this tiny shrew, which weighs just seven-hundredths of an ounce, the weight of a pin, is actually a giant glutton? It's true: The pygmy shrew eats more than twice its weight every day. And some of its prey is as big as it is. To overcome insects, the shrew bites them, injecting them with a paralyzing venom. The shrew has to eat so much to keep up its strength because it is constantly on the move.

# What is the most isolated land on earth?

> The island of Tristan da Cunha.

Tristan da Cunha, an island in the Atlantic Ocean, is the most remote inhabited land in the world. The closest land inhabited by people is the island of Saint Helena, 1,510 miles away. Discovered in 1506, the island was not visited for years except by whale hunters. In 1806, a few brave souls unafraid of a solitary life moved there. Between 1908 and 1918, the island's inhabitants received not a single visit. Today, only 270 people live there. There is no airport, so a boat comes to bring them provisions once every six weeks.

# How tall is this prehistoric mushroom?

**> It can grow up to twenty feet tall.**

This mushroom would never have fit in a frying pan. Prototaxites (pronounced *pro-toe-TAX-it-eez*), a mushroom that grew four hundred million years ago, was twenty feet tall—as tall as two basketball hoops stacked on top of each other. In recent years, prototaxites fossils have been discovered all over the world. For a long time, it was thought they were trees or a type of algae. It was only in 2007 that scientists declared that prototaxites had actually been giant mushrooms.

**A reconstruction of prototaxites as they might have appeared.**

# Does the millipede really have 1,000 feet?

> No, but it has more than 750.

The millipede doesn't exactly deserve its name: There are no millipedes with 1,000 feet. But it certainly has a lot. The record is held by a rare species of millipede with 750 feet, discovered in a California forest in 1926. It was believed this species had since disappeared until a scientist found some of its descendants in the same forest in 2005.

# What volcano is covered with snow?

> Mount Erebus, in Antarctica.

Hidden under the snows of the South Pole is Mount Erebus, an active volcano. When you look at a globe, this is the volcano closest to the bottom. Another of its particularities is that it's so well covered by snow, we don't know much about it. In this part of the world, the weather is so extreme (the temperature ranges from –58 degrees to –4 degrees Fahrenheit) that scientists are only able to climb up Mount Erebus to study it one month a year.

# What makes as much noise as a jackhammer?

> A male African cicada.

The male African cicada is the loudest insect in the world. How does it make so much noise? By vibrating two membranes under its abdomen. And what's the point of making such a racket? To attract females, which are themselves silent, and to scare away predators.

# Contents

# photo credits

pp. 10–11: All rights reserved
pp. 12–13: All rights reserved
pp. 14–15: All rights reserved
pp. 16–17: All rights reserved
pp. 18–19: All rights reserved
pp. 20–21: All rights reserved
pp. 22–23: All rights reserved
pp. 24–25: © Bob Rowan; Progressive Image/CORBIS
pp. 26–27: All rights reserved
pp. 28–29: All rights reserved
pp. 30–31: © Rick Price/CORBIS
pp. 32–33: All rights reserved
pp. 34–35: All rights reserved
pp. 36–37: All rights reserved
pp. 38–39: All rights reserved
pp. 40–41: All rights reserved
pp. 42–43: All rights reserved
pp. 44–45: All rights reserved
pp. 46–47: © JOUAN Catherine/RIUS Jeanne/Jacana/Eyedea Illustration
pp. 48–49: All rights reserved
pp. 50–51: All rights reserved
pp. 52–53: © WALKER Tom/JACANA/Hoa-qui/Eyedea Illustration
pp. 54–55: All rights reserved
pp. 56–57: © Douglas Peebles/CORBIS
pp. 58–59: All rights reserved
pp. 60–61: All rights reserved
pp. 62–63: All rights reserved
pp. 64–65: All rights reserved
pp. 66–67: All rights reserved
pp. 68–69: All rights reserved
pp. 70–71: © Bob Gomel/CORBIS
pp. 72–73: © Bruno Cossa/Grand tour/CORBIS
pp. 74–75: All rights reserved
pp. 76–77: All rights reserved
pp. 78–79: All rights reserved
pp. 80–81: All rights reserved
pp. 82–83: All rights reserved
pp. 84–85: All rights reserved
pp. 86–87: All rights reserved
pp. 88–89: All rights reserved
pp. 90–91: All rights reserved
pp. 92–93: All rights reserved +
pp. 94–95: All rights reserved
pp. 96–97: © Ronald Wittek/dpa/CORBIS
pp. 98–99: All rights reserved
pp. 100–101: All rights reserved
pp. 102–103: All rights reserved
pp. 104–105: © JOHN DOWNER/NPL/Jacana/Eyedea Illustration
pp. 106–107: All rights reserved
pp. 108–109: © NASA/4646/Gamma/Eyedea Presse
pp. 110–111: © Macduff Everton/CORBIS
pp. 112–113: All rights reserved
pp. 114–115: © SAN BUGHET/Hoa-qui/Eyedea Illustration
pp. 116–117: All rights reserved
pp. 118–119: All rights reserved
pp. 120–121: All rights reserved
pp. 122–123: © All rights reserved
pp. 124–125: All rights reserved
pp. 126–127: All rights reserved
pp. 128–129: All rights reserved
pp. 130–131: All rights reserved
pp. 132–133: All rights reserved
pp. 134–135: All rights reserved
pp. 136–137: All rights reserved
pp. 138–139: © Mike Theiss/Ultimate Chase/CORBIS
pp. 140–141: All rights reserved
pp. 142–143: All rights reserved
pp. 144–145: All rights reserved
pp. 146–147: All rights reserved
pp. 148–149: All rights reserved
pp. 150–151: All rights reserved
pp. 152–153: All rights reserved
pp. 154–155: All rights reserved
pp. 156–157: All rights reserved
pp. 158–159: All rights reserved
pp. 160–161: All rights reserved
pp. 162–163: © Arctic-Images/CORBIS
pp. 164–165: © Ruffier-Lanche/Jacana/Eyedea; All rights reserved
pp. 166–167: © DOC WHITE/NPL/Jacana/Eyedea Illustration
pp. 168–169: All rights reserved
pp. 170–171: All rights reserved
pp. 172–173: © All rights reserved
pp. 174–175: All rights reserved
pp. 176–177: All rights reserved
pp. 178–179: All rights reserved
pp. 180–181: © B.Borell Casals; Frank Lane Picture Agency/CORBIS
pp. 182–183: © Delphine Aure/Jacana/Eyedea
pp. 184–185: All rights reserved